Clifton Park - Halfmoon Public Library

Guess Who
Snaps

Sharon Gordon

BENCHMARK BOOKS

MARSHALL CAVENDISH
NEW YORK

Can you see me in the water?

I am hiding.

I stay in my *den* during the day.

I walk on the bottom of the sea at night.

My eyes are on *stalks*.

They move all around.

I have eight legs.

If I lose a leg, I can grow
a new one.

I have a hard shell.

It keeps my body safe.

My shell is green and black on top.

It is orange on the bottom.

As I grow, my shell gets too tight.

I *shed* my old shell.

Sometimes I eat it!

I like to eat small fish and animals.

I catch them with my claws. *Snap!*

I use one claw to rip.

I use one claw to crush.

I lay thousands of eggs.

They hatch in ten months.

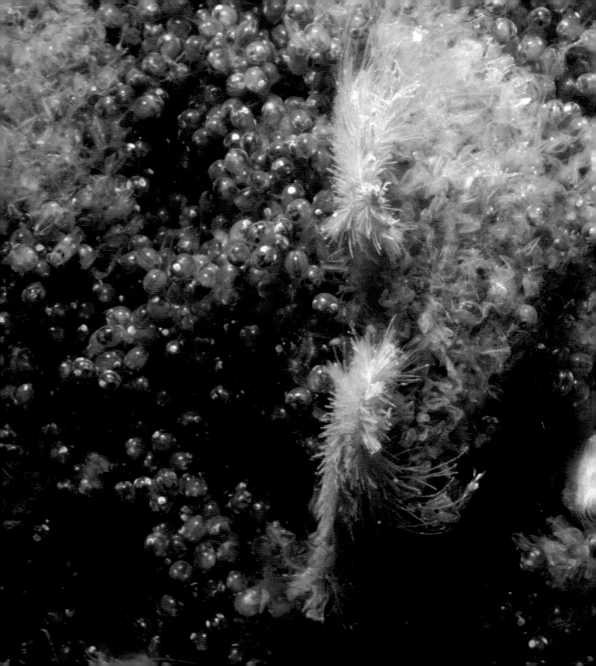

My babies look like tiny bugs.

Only a few will grow up.

Do you see me in the tank?

Watch your fingers!

Who am I?

I am a lobster!

Who am I?

claws

eggs

eyes

legs

shell

Challenge Words

den
A lobster's home.

shed
To fall off or lose.

stalks
The short rods that hold a lobster's eyes.

Index

Page numbers in **boldface** are illustrations.

About the Author

Sharon Gordon has written many books for young children. She has always worked as an editor. Sharon and her husband Bruce have three children, Douglas, Katie, and Laura, and one spoiled pooch, Samantha. They live in Midland Park, New Jersey.

With thanks to Nanci Vargus, Ed.D. and Beth Walker Gambro, reading consultants

Benchmark Books
Marshall Cavendish
99 White Plains Road
Tarrytown, New York 10591-9001
www.marshallcavendish.com

Library of Congress Cataloging-in-Publication Data

Gordon, Sharon.
Guess who snaps / by Sharon Gordon.
p. cm. — (Bookworms: Guess who)
Includes index.
ISBN 0-7614-1765-6
1. Lobsters—Juvenile literature. I. Title II. Series: Gordon, Sharon. Bookworms: Guess who.

QL444.M33G67 2004
595.3'84—dc22
2004006872

Photo Research by Anne Burns Images

Cover Photo by Animals Animals/Carmela Leszczynski

The photographs in this book are used with permission and through the courtesy of:
Peter Arnold, Inc.: pp. 1, 7, 28 (bottom l.) Kelvin Aitken; p. 5 Fred Bavendom;
pp. 9, 28 (bottom r.) Norbert Wu; pp. 21, 28 (top r.) Jeffrey L. Rotman. Jonathan Bird, 1995:
p. 3. Photo Researchers: p. 23 George Lepp; p. 27 Andrew Martinez. Corbis: pp. 11, 13, 29
Stephen Frink; p. 25 Judy Griesedieck. Norbert Wu Productions: p. 15 Mark Conlin.
Animals Animals: p. 17 Carmela Leszczynski; pp. 19, 28 (top l.) Herb Segars.

Series design by Becky Terhune

Printed in China
1 3 5 6 4 2

FEB   2008